Introduction

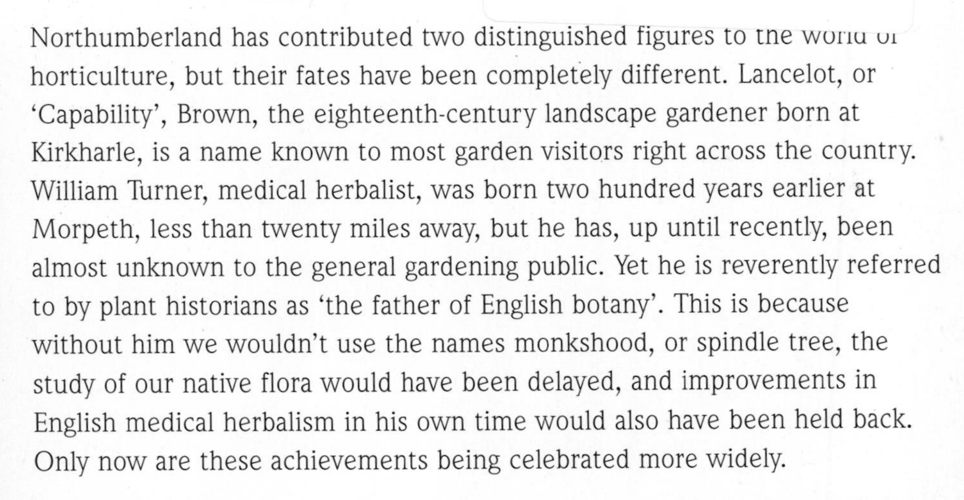

Northumberland has contributed two distinguished figures to the world of horticulture, but their fates have been completely different. Lancelot, or 'Capability', Brown, the eighteenth-century landscape gardener born at Kirkharle, is a name known to most garden visitors right across the country. William Turner, medical herbalist, was born two hundred years earlier at Morpeth, less than twenty miles away, but he has, up until recently, been almost unknown to the general gardening public. Yet he is reverently referred to by plant historians as 'the father of English botany'. This is because without him we wouldn't use the names monkshood, or spindle tree, the study of our native flora would have been delayed, and improvements in English medical herbalism in his own time would also have been held back. Only now are these achievements being celebrated more widely.

In his own day Turner was also famous - or notorious - for his fierce religious convictions, which impinged on his fortunes, and therefore his career, throughout his life. So in order to understand how he came to acquire his reputation at the head of English botany, we need to look briefly at Turner's life and times, then to consider the state of medical botany in his era and finally to find out exactly how he contributed so outstandingly to improving his countrymen's knowledge of native and foreign plants.

Aconitum napellus, given the English name 'monkshood' by Turner.
(photograph: Rachel Singleton)

William Turner - Father of English Botany
by Marie Addyman
First Published 2008 by Friends of Carlisle Park,
Morpeth, England
www.focpMorpeth.org
ISBN: 978-4-907114-00-7

Henry VIII, in whose reign Turner began his studies and medical career.
(courtesy of the National Portrait Gallery, London)

Turner's life and times

There are no painted likenesses extant of Turner; instead, there are numerous comments on his character. What comes across most clearly is that he always did what he believed in - whether this was writing a herbal in English, or criticising bishops for their regalia. To those whom he opposed and who opposed him, he seemed troublesome and irascible. To those who worked with him, he was appreciated as a man of unflinching integrity.

Certainly Turner was a brave man, for he stuck to his principles while living in a most unsettled period of English history: the reigns of Henry VIII, and his children Edward VI, Mary, and Elizabeth. During those reigns his works were banned by Henry and Mary, he was exiled by Henry and Mary, he seems to have been imprisoned under Henry and he was suspended for nonconformity under Elizabeth. Two brief periods of comparative peace saw him employed in the reign of Edward by Protector Somerset at Syon House in London from 1547-49; and his appointment as Dean of Wells, made in 1551, eventually gave him a few years of peace at the very end of his life in the 1560s.

Turner contributed his knowledge of birds and fishes to his European colleagues. A 16th C. woodcut of the stork.

(reproduced in a 19th C. volume on Shakespeare's natural history by H. W Seager)

The daffodil, carefully researched by Turner, and portrayed in an early Tudor text.

(MS Ashmole 1504 fol.9v, courtesy of the Bodleian Library, University of Oxford)

Early life in Morpeth

William Turner was born in Morpeth. That much is known for certain, but other early details are speculative. It's thought that he was born about 1508 or perhaps a year or two later, it's believed that his father ran a prosperous tanning concern in the town and it's assumed that he attended the local school beside the bridge. This was a chantry school, a religious endowment, which would give him a good basic education. Whether Turner was able to return home frequently or at all in later life is dubious, but he undoubtedly held his birth-place in great affection. In his will he remembered residents of both Morpeth and Hexham. His writings are full of references to Morpeth, the surrounding area, and the coastal strip from Tynemouth to Lindisfarne.

The robin, distinguished by Turner from the redstart, in defiance of received classical wisdom. (painting by John Caffrey)

Most important for us, it was in his boyhood that Turner began that process of observing the natural world which was to be a key element in his published writings. He observed not only plants of all kinds - trees, ferns, grasses and wild flowers - but living creatures too - fishes and birds. His observations of fishes were so accurate and detailed that his notes would be incorporated into the work of the great naturalist Conrad Gesner, whom he would meet in the 1540s. During the same decade, his own published work on birds would be full of those early Northumbrian memories. In one of the most famous entries he describes a robin's nest 'built at the root of brambles where oak leaves lie thick', its entrance carefully camouflaged. Later, when he was studying Aristotle and read of the great master's conflation of a robin with a redstart, it would be early observation like this which would enable him to see that classical authority and local knowledge did not always match up. All-in-all, it's thought that Turner was responsible for 37 first records of English birds. With so much empirical investigation already achieved, and with some sound book-learning - probably including Latin - drilled into him, Turner went up to Pembroke Hall, Cambridge as an undergraduate in 1526.

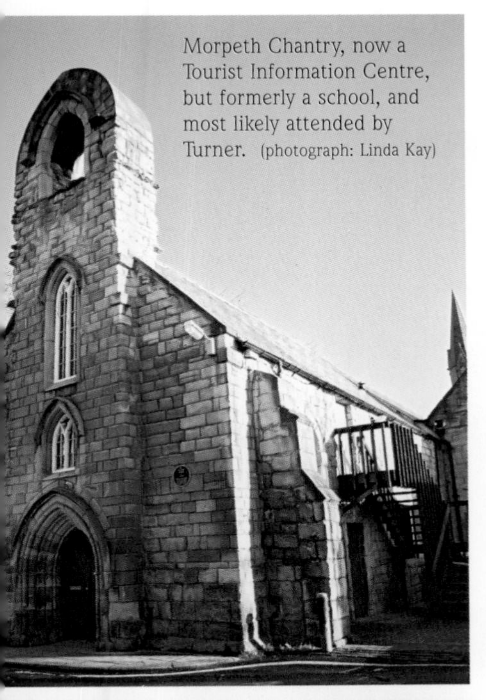

Morpeth Chantry, now a Tourist Information Centre, but formerly a school, and most likely attended by Turner. (photograph: Linda Kay)

Turner's student years

Referring back to those student years, Turner described himself as a student 'of Physic and philosophy' (1562), a combined programme of study which is slightly confusing to the modern reader. But the university syllabus common to both Oxford and Cambridge, and more or less shared throughout European universities, was very different in the sixteenth century. Students who followed the degree programme were committed to a seven-year period of study to gain an MA, with the interim BA awarded after four years. This BA-MA was still in essence a general degree, based on the mediaeval curriculum and including what we would think of as both arts and sciences. Only after the completion of this course of work did the student go on to the specialist areas of theology, law, or physic. Turner followed this standard programme, through BA to MA. His specialist degrees in medicine took longer to acquire over the troubled period of his first exile (discussed below), but in the 1540s he earned his MD from Ferrara and then from Oxford on his return to England.

Turner's study of 'physic' as a career choice is easily understood. His reference to 'philosophy' might suggest that he developed his interest in 'natural philosophy' - what we would call natural history, or perhaps natural science - as an extension of his studies in physic. Certainly he took advantage of the opportunities being offered at Cambridge. At this period the Oxbridge colleges were increasingly going beyond their remit to provide housing and pastoral care for students, and offering actual tuition of a kind which interpreted the official syllabus in a more flexible way. One example of this flexibility, from which Turner was to benefit, was that the call by Erasmus to study the Greek originals of the key texts in the curriculum was being met in some parts of Cambridge with great enthusiasm.

Left: The remains of Our Lady's Chapel, Bothal Woods, which is where Turner records finding broomrape growing. (photograph: A.W. Davison)

Links with Northumberland

As the son of an artisan, however prosperous, Turner would be in a minority while up at Cambridge. However, as the recipient of a scholarship provided by Thomas, first Lord Wentworth, he would be part of a reputable and flourishing tradition. Several of the Oxbridge colleges had patronage links with individual counties, sending out agents to recruit bright local youths from the schools. These students would be educated and then sent back as minor officials to their own county. Turner was atypical in not following the civil service route, but typical in being selected for a college with strong links to his own county. Northumbrians were visible throughout the whole hierarchy of Pembroke College, supplying college masters, lecturers and students. It was there that Turner met distinguished Northumbrians, including Nicholas Ridley, who taught him Greek. He also met other soon-to-be-famous men like William Burghley, future statesman under both Edward and Elizabeth, and Roger Ascham, Elizabeth's future tutor.

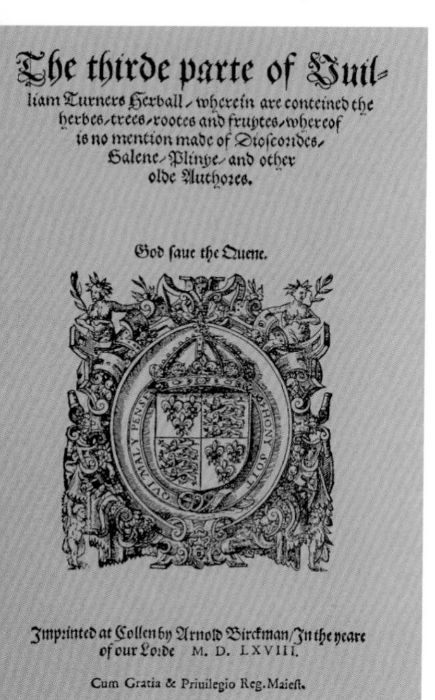

The thirde parte of Uuilliam Turners Herball, wherein are conteined the herbes, trees, rootes and fruytes, whereof is no mention made of Dioscorides, Galene, Plinye, and other olde Authores.

God saue the Quene.

Imprinted at Collen by Arnold Birckman, In the yeare of our Lorde M. D. LXVIII.

Cum Gratia & Priuilegio Reg. Maiest.

Left: William Turner's dedication to Queen Elizabeth I, from the frontispiece to the third part of his *Herball* of 1568. (Castle Morpeth Borough Council)

Elizabeth I, in whose reign Turner finally achieved some stability in which he was able to complete his *Herball*. (courtesy of the National Portrait Gallery, London)

Willimotswick, a fortified farmhouse belonging to the Ridley family in the Tyne valley.
(photograph: Linda Kay)

Religious upheaval

Ridley taught Turner far more than Greek. He and his future co-martyr under Mary, Hugh Latimer, became key voices in the 1530s in proclaiming the need for religious reformation in England, a cause which Turner enthusiastically espoused. The 1530s was a troublesome and an exciting decade for England, for Cambridge, and for Turner. It saw Henry VIII gain his divorce, and assert his independence of Papal authority in England. But in declaring himself to be supreme Head of the Church in England, Henry was not declaring for the Protestantism being canvassed in Europe by Luther, Zwingli or Calvin. Though he may have destroyed the monasteries, he was still deeply committed to both Catholic doctrines and to a Catholic organisation of the Church, with its bishops, its hierarchy, and its full ceremony and regalia. This brought him into conflict with those who, like the Cambridge group, really wanted to embrace continental Protestantism - to do away with the Mass and most of the sacraments, to abolish bishops and ceremonial, and to make the services, the prayer book, and the Bible available in English. In short, while he was denouncing Roman Catholic authority in the person of the Pope, Henry was enforcing Catholic doctrines, and persecuting those 'Protestants' who advocated otherwise.

Religion and medicine

The religious politics of the 1530s are the forum for Turner's activities after graduating, and for his exile at the end of the decade. They establish the pattern of years to come, showing his religious beliefs and his medical commitments to be inseparable in both his thoughts and in his career. His first written works took the form of a religious tract and a medico-botanical one, the *Libellus de re herbaria* (1538) - a pairing which would continue all his life. For Turner, religion and physic were not separate activities. Physic is the compassionate activity 'openly commended' as a 'singular gift of God' in the Bible (1551). Its language, for Turner as for Shakespeare, offers the metaphor of purging the corrupt diseases with which the body politic is riddled. Truth-telling, condemning abuse, is 'spirituall Physik' (1555); it provides a 'triacle' against spiritual poison (1552) - the old word to describe a carefully concocted antidote against the most lethal poisons

Troubled times

Turner was fully supported by his College in these beliefs and practices. After acquiring his BA (1529/30), he became a Fellow, and went on to gain his MA (1533); he was ordained deacon from Lincoln, then the diocesan cathedral for Cambridge (1537), and he claimed he had also received license to preach required at the time; later, he would be offered financial support while in exile. But outside this College world of like-minded folk, he fell foul of the dangerous world of Henry VIII's newly created national Church. He was ordained within that state Church, while a Protestant in his personal beliefs, so the preaching career he embarked on round Oxford would probably run contrary to current orthodoxy - it is therefore possible that he was imprisoned at this point. Also, he had decided to marry Jane Auder, the daughter of a Cambridge alderman - an action in full conformity with Lutheran belief in a married priesthood, but totally contrary to Henry's strict ordinances, which enforced celibacy on anyone who had received minor orders, such as that of deacon. With some friends being persecuted and others executed, Turner ended the decade by fleeing to the Continent.

The original entrance to the Physic Garden at Padua, established in the 1540s.
(photograph: Marie Addyman)

Right: *Eryngium maritimum*, referred to in herbals from Dioscorides to Turner, grows wild on the Northumberland coast. This specimen of sea-holly however is in the Oxford Botanic Garden.
(photograph: Marie Addyman)

Below: Illustration of *Eryngium fuchsii*, from Turner's *Herball* of 1551.
(Castle Morpeth Borough Council)

Travels and medicinal herbalism

The importance of Turner's travels abroad for his work in medical herbalism will be discussed subsequently, but we can't help but wonder how he reconciled his austere Protestantism with the flamboyance of Catholic Italy. In fact, Turner spoke of his time there, of the colleagues he worked with, and particularly of Luca Ghini of Bologna, with affectionate respect. And in one instance at least, his respect for good scholarship forced a grudging acknowledgment that sometimes botany and religious politics just had to be considered separately. Trying to distinguish 'borage' from the 'buglossum' of the classical Greek herbalist Dioscorides, he cites 'two friars of Rome' who had been researching plants in Spain. After a sustained analysis of available plants and available texts, he concludes:

> 'Let learned men judge both by… the judgement of the two friars of Rome, whom I cannot so much dispraise for the hypocritical kind of living, being in Babylon, as I can allow them for their diligent labours taken in seeking out of simples, and restoring Mesue unto his right and true text, and first writing.' (1551)

The mulberry tree (*Morus nigra*) which grows in the garden at Syon House is popularly believed to have been planted by Turner.
(MS Ashmole 1504, fol.22v, courtesy of the Bodleian Library, University of Oxford)

Political changes

While abroad, Turner continued both his religious polemic, being able to criticise freely and bluntly from the safety of the Continent, and his botanical researches. He studied various plants and habitats as he travelled through Germany, and met the great Conrad Gesner, comparing notes with him in Zurich on various aspects of natural history and contributing information and observation about fishes to Gesner's encyclopaedic work. It's probable that he supported himself and his family by working as a physician, since he ended up spending four years in that capacity for the Duke of Emden, in low-lying, marshy East Friesland. This gave him the opportunity to study an entirely different habitat and entirely different plants. It may have been during this period that he was oscillating between writing in Latin for the international community, or in English for his fellow-countrymen back home

His decision must have been made by the time he returned to England in 1547, since *The Names of Herbs,* written in English, followed quite soon afterwards. Turner was able to return home because the death of Henry brought to the throne his son Edward VI, a minor legally under the Protectorship of Edward Seymour, Duke of Somerset, a committed friend of Protestantism. He was also related to the Wentworths, so the old connection to his university patron may have got Turner the post of physician in Somerset's household. For two years he was in a milieu which was at the centre of English politics, himself becoming an MP in 1547, and able to see the long-desired Reformation finally taking shape.

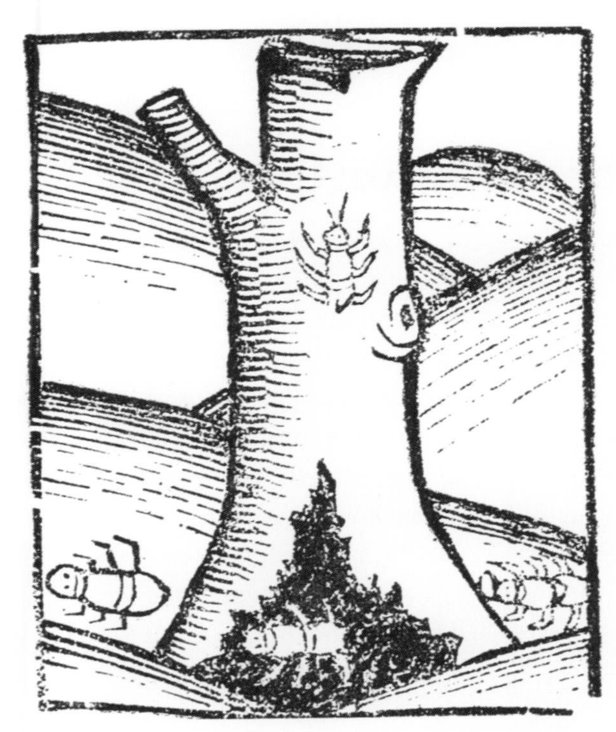

Gardeners in the 16th century distrusted all creeping and flying creatures, from ants to worms and scorpions (the latter very common, of course, in English gardens!).

(A 16th C. woodcut of the ant at work, reproduced in a 19th C. volume on Shakespeare's natural history by H. W Seager)

The gallows at Tyburn, a place of dread throughout Turner's lifetime, but still a village in the 16th century, and site of several native wild flowers.

A peaceful interlude

For a brief time, he was able to write in comparative peace, and to botanise in another area of England, since Somerset's main London residence was at Syon House, near the Thames. Turner supervised the garden there - an old tradition says the mulberry still growing there was planted by him - and he probably had his own garden attached to his residence at Kew. Presumably he also had some access to the gardens of the Protector's wealthy associates. Hence he was able to see some of the brilliant new plants entering horticulture, and also to scour the banks of the Thames for wild plants. Since London was still a series of villages at this time, native plants abounded. John Gerard, writing fifty years later, still finds bugloss in 'Piccadilla', and mallow at 'the place of execution called Tyburn'.

Turner does not seem to have felt very secure financially while working for Somerset, for he was even then trying to gain other employment, either academic or ecclesiastical. But this cannot compare with the desperate situation he faced with the fall of his patron in 1549. The new Protector, Northumberland, was certainly not a man with a feeling for Turner's old Northumbrian connections. The next few years saw Turner appealing to William Cecil, or to anyone who could assist him, until he finally acquired what was probably an ecclesiastical sinecure, the Prebendary of Botevant in York, since there is no record of his attendance there.

Turner's *Herball* (part I) published

The most important appointment, however, was as Dean of Wells, in 1551. This should have brought security, but the post was already occupied by John Goodman, a clergyman who adhered to Catholic forms and beliefs. His superior, Bishop William Barlow, a married clergyman who embraced the new Protestantism, wanted to eject him, but Goodman hung on, taking his case to court and refusing to relinquish his house. Turner, meanwhile, ill and frustrated, lodged with his family in great discomfort, yet he pushed on with the publication of the first part of his *Herball* – dedicated, in a gesture of loyalty, to his old master Somerset. It also seems likely that he went to work immediately on the second part, though circumstances would prevent this being published until 1562.

Wells Cathedral.
(courtesy of the Chapter of Wells, by kind permission)

Although Turner finally got access to his house in Wells at the end of 1552, by the following year he and other-fellow Protestants were in exile again. The accession of Mary, determined to return England to the Roman Catholic fold of her childhood, resulted in the exile or execution of hundreds of her subjects. Returning to Germany with his family, Turner again worked as a physician, again botanised whenever he had the chance, and again sent furious pamphlets across the Channel, lambasting the heresies of Mary and the Bishop of Winchester, Stephen Gardiner, his old enemy the 'Romish foxe' who had now re-emerged even more virulently as a 'Romish wolfe'.

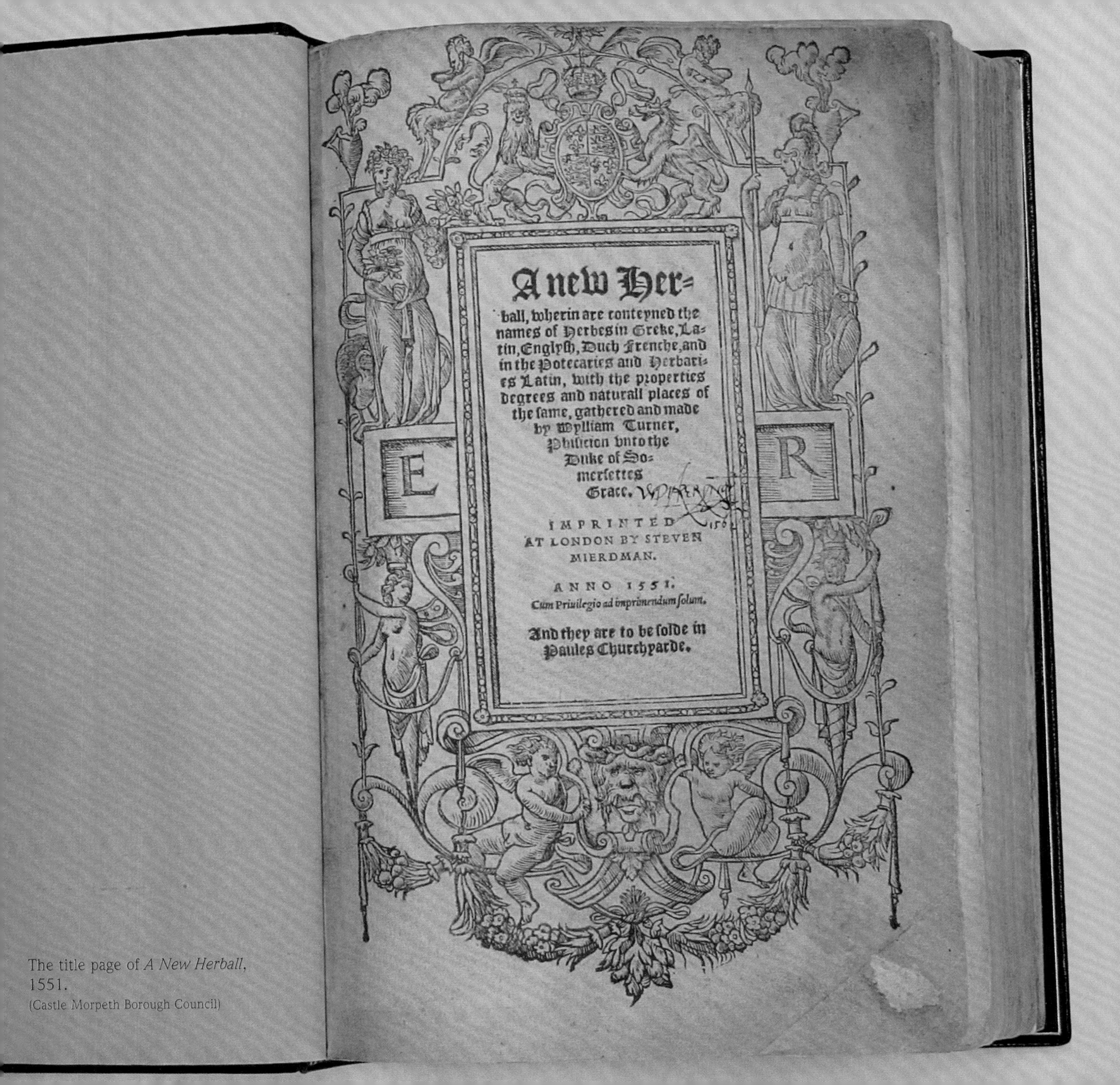

A new Her-
ball, wherin are conteyned the
names of Herbes in Greke, Latin, Englysh, Duch Frenche, and
in the Potecaries and Herbaries Latin, with the properties
degrees and naturall places of
the same, gathered and made
by Wylliam Turner,
Phisicion vnto the
Duke of So-
mersettes
Grace.

IMPRINTED
AT LONDON BY STEVEN
MIERDMAN.

ANNO 1551.
Cum Priuilegio ad imprimendum solum.

And they are to be solde in
Paules Churchyarde.

The title page of *A New Herball*, 1551.
(Castle Morpeth Borough Council)

Elizabeth I rules England

Only the accession of Elizabeth in 1558 finally seemed to promise peace and prosperity, though the results were not immediate. England was now awash with former exiles whose loyalty to Elizabeth was unquestioned, but who returned to claim back ecclesiastical positions and properties taken over by others in Mary's reign. Moreover, Turner and his fellows had again spent time in the fervent Protestant countries of Switzerland, Germany, and Holland. They were returning, as they hoped, to see that incipient Reformation begun under Edward confirmed under his sister, while Elizabeth herself intended to proceed cautiously in the pursuit of the famous 'middle way' that would become the Anglican Church. Turner found that both his house and his position at Wells had been re-occupied, and that the current Bishop of Wells adhered to those forms of ceremonial and regalia which he detested. Hence he lived in London on his return; only in 1560 did he gain access to his house at Wells.

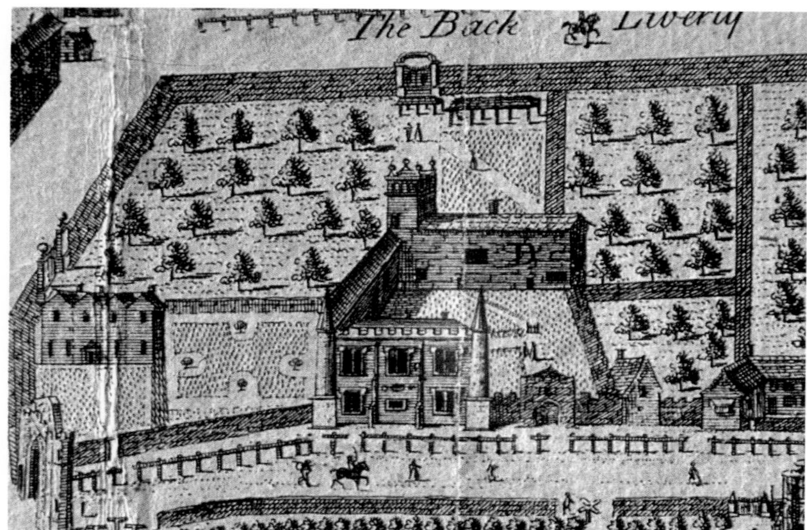

A period of teaching and study

These are the years when Turner involves himself by example, by teaching and by pamphlet in the Vestiarian Controversy - the debate over clerical dress and regalia.
He must have been a thorn in his bishop's side, notoriously teaching his dog to steal his official headgear, and ordering a 'common adulterer' to do public penance while wearing a priest's cap. Somehow he survived the antagonism he caused, as well as the severe illness which drove him back to London in 1566.

Left: This picture map, drawn by William Simes in 1735, is the earliest plan of the Deanery Garden.
(courtesy Richard Neale OBE)

Turner's *Herball* (parts II and III) published

Booke of the natures and
erties, as well of the bathes in England as of other bathes in Germanye and Italye, very necessarye
for all sycke persones that can not be healed
without the helpe of natural bathes,
lately ouersene and enlarged by
William Turner Doctor
in Physick.

God saue the Quene.

Imprinted at Collen by Arnold Birckman/In the yeare
of our Lorde M. D. LXVIII.

Cum Gratia & Priuilegio Reg.Maiest.

Above: Turner's *Book of the Natures and ... bathes*, written at Wells in the 1560s, showing that his medical botany was part of a more general interest in health issues.
(courtesy the Wellcome Library, London)

Turner worked steadily, completing the long, and much interrupted, second part of his *Herball* in 1562, gratefully dedicated to Sir Thomas Wentworth, son of the man who had sent him to university nearly thirty years beforehand. He continued studying the plants around Wells, and he completed his life's work by producing the third and final part of his great work, describing those plants that had never been included by Dioscorides or other classical writers, in 1568.

This final work, like its previous counterparts, was still sharply focussed on the practical needs of his fellow-countrymen: it was dedicated to the surgeons of London, and it gave them technical information about the degrees of herbs, as described in the next section. With the same sense of his colleagues' needs in mind, he had already published with his second volume, a new edition of an earlier tract he'd written on the healthful properties of spas and baths. He also corrected an earlier work on 'triacles' and antidotes to poisons, co-published with a 'new book' on the nature and properties of wines. In this way, Turner situated his work on medical herbalism within a broader programme of overall health-care.

Unremitting in his scholarship and aided now by his son, he corrected earlier mistakes and expanded earlier information. He must have laboured on right up to the very end, for all these works were published in 1568, the year he died at his house in Crutched Friars, London. He was buried at St Olave's Church. His widow married another Marian exile, Richard Cox, Bishop of Ely. Turner's son who had worked with him on correcting his *Herball*, carried on the family tradition, himself becoming a distinguished physician.

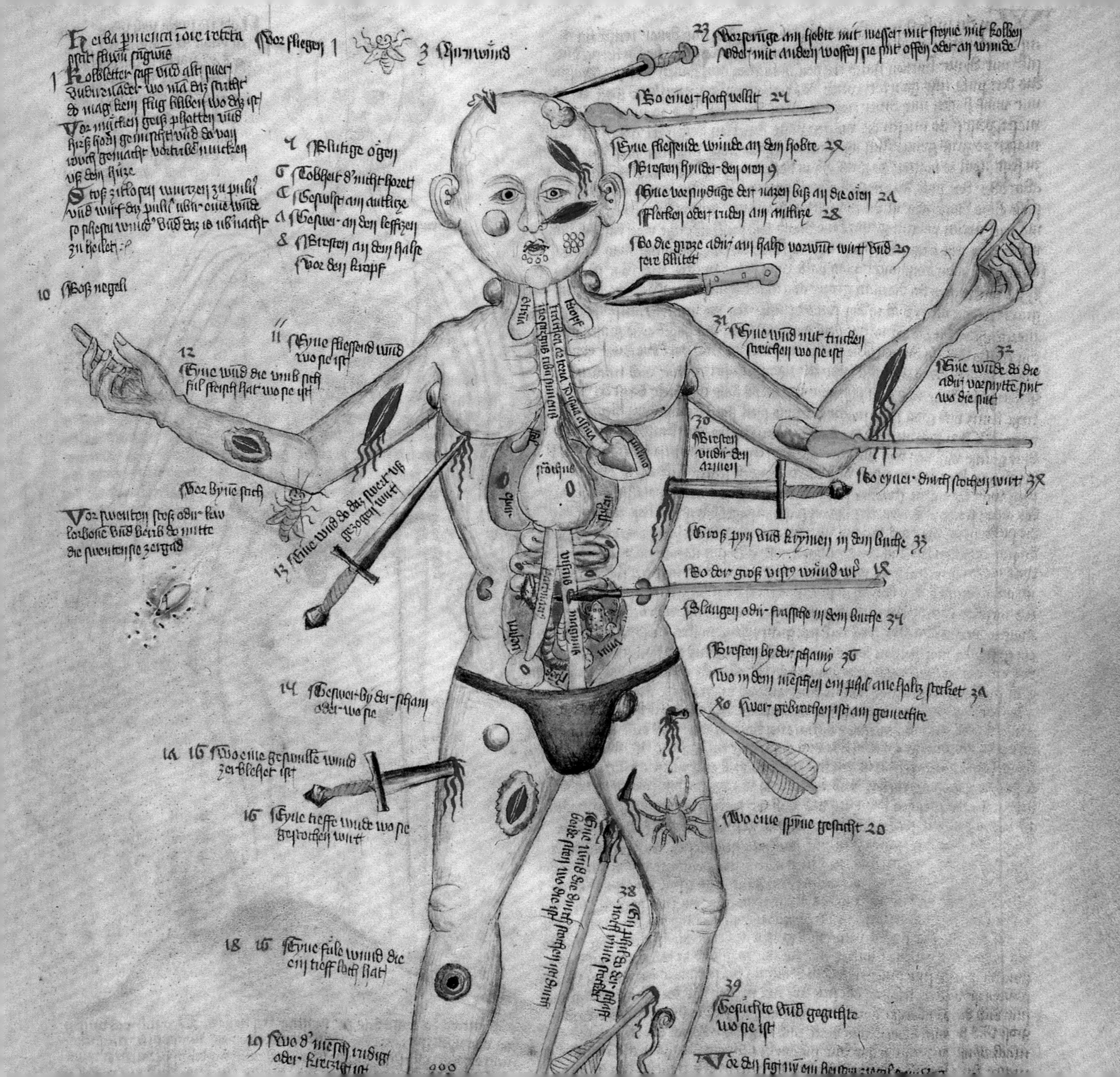

Vor fliegen 1
Blutige ogen
Boß negel
Eyne fliessende wund wo sie ist
Vor byne stich
Eyne fliessende wunde an dem hobte 25
Bresten hinder den oren
Flecken oder ruden am antlitze 28
Bresten by der schame
Wo eine spynne gestochen 20
Eyne tieffe wunde wo sie gestochen wirt
Eyne fule wund die ein tieff loch hat

Sixteenth century medicine and medical botany

When Turner left England in 1540, he probably left his wife in Germany, but he himself headed down to Northern Italy - to Venice, Ferrara and Bologna. His strong religious convictions would have attracted him to Protestant Germany, but Turner was also in the process of completing his medical training, and for this the cities of Northern Italy served him far better.

New medical practice in Italy

Throughout the sixteenth century, Padua (the official university of the Venetian republic), Ferrara, Bologna and Pisa were academic centres of excellence to which aspiring English physicians would gravitate. The 1540s in particular were exciting times in Italian medical departments. At Padua, surgery was included in the medical course, and the brilliant young anatomist Andreas Vesalius was making his mark with his dynamic style of lecturing. Clinical practice - the actual teaching of students at patients' bedsides - was being introduced and Padua was vying with Pisa to create the first physic garden in Europe, forerunner of all future botanic gardens. Meanwhile at Bologna, the great medical botanist Luca Ghini, whom Turner cites repeatedly as his 'master' and with whom he studied, was corresponding with fellow-botanists across Europe to identify names correctly, before becoming the first director of the new garden at Pisa.

This was in strong contrast to the medical departments at Oxford and Cambridge, where the syllabus was old-fashioned, and where there were too few students to provide a fully trained corpus of professional physicians. Thomas Linacre had tried to rectify the situation in 1517 by persuading the king to set up a Royal College of Physicians. In 1523, only twelve Fellows were recorded, with perhaps another six by the end of the 1530s. To be as up-to-date as possible in medicine therefore, it was better to obtain qualifications in Northern Italy.

Above: England had to wait until 1621 before following the example of Italian universities and setting up a physic garden. Oxford Botanic Garden (renamed in 1840) still retains its original entrance and pattern of formal beds.
(photographs: Marie Addyman)

Left: A mediaeval image of 'Wound Man', showing the different ways he could be wounded and what were the effects.
(courtesy of The Wellcome Library, London)

Wise women and apothecaries

The university-trained physician of the mid-sixteenth century was at the top of a complicated hierarchy inherited from the Middle Ages. At the bottom were those who provided the largest service across the country: women - whether wise women in the village community, ladies of the manor administering to the household and the wider community or women who set up as midwives. They would be the mainstay of the poor, both urban and rural. Some of these women would collect 'simples', individual plants which they sold to the apothecaries. These, the forerunners of our chemists, made up prescriptions according to the directions of those ranked above them. They were not supposed to prescribe but they often did, like Nicolas Culpeper, the mid-seventeenth-century herbalist for London's poor.

Woman de-lousing a man's head.
(16th C. woodcut reproduced in a 19th C. volume on Shakespeare's natural history by H. W Seager)

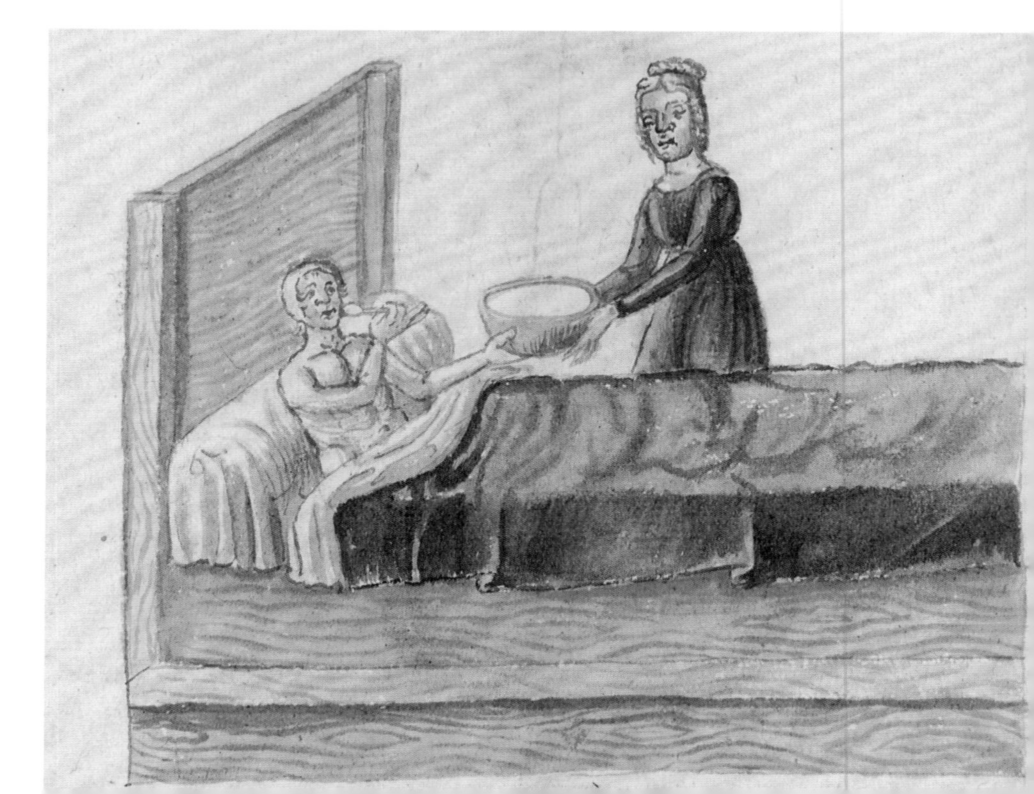

Surgeons

Above the apothecaries were the surgeons, who defended their right to deal with any procedure involving the knife - cauterising, amputating, operating, pulling teeth. Some universities did offer specific courses in surgery, or some surgeons might follow part of a university medical course, but in theory the surgeon was trained by the guild, rather than the university. In practice the supposed difference in knowledge was sufficiently blurred by the end of Turner's life for him to dedicate the third part of his *Herball* to the Company of London Surgeons.

Left: Woman tending the sick.
(courtesy of the Wellcome Library, London)

Right: A late mediaeval image of man and the macrocosm.
(courtesy of the Wellcome Library, London)

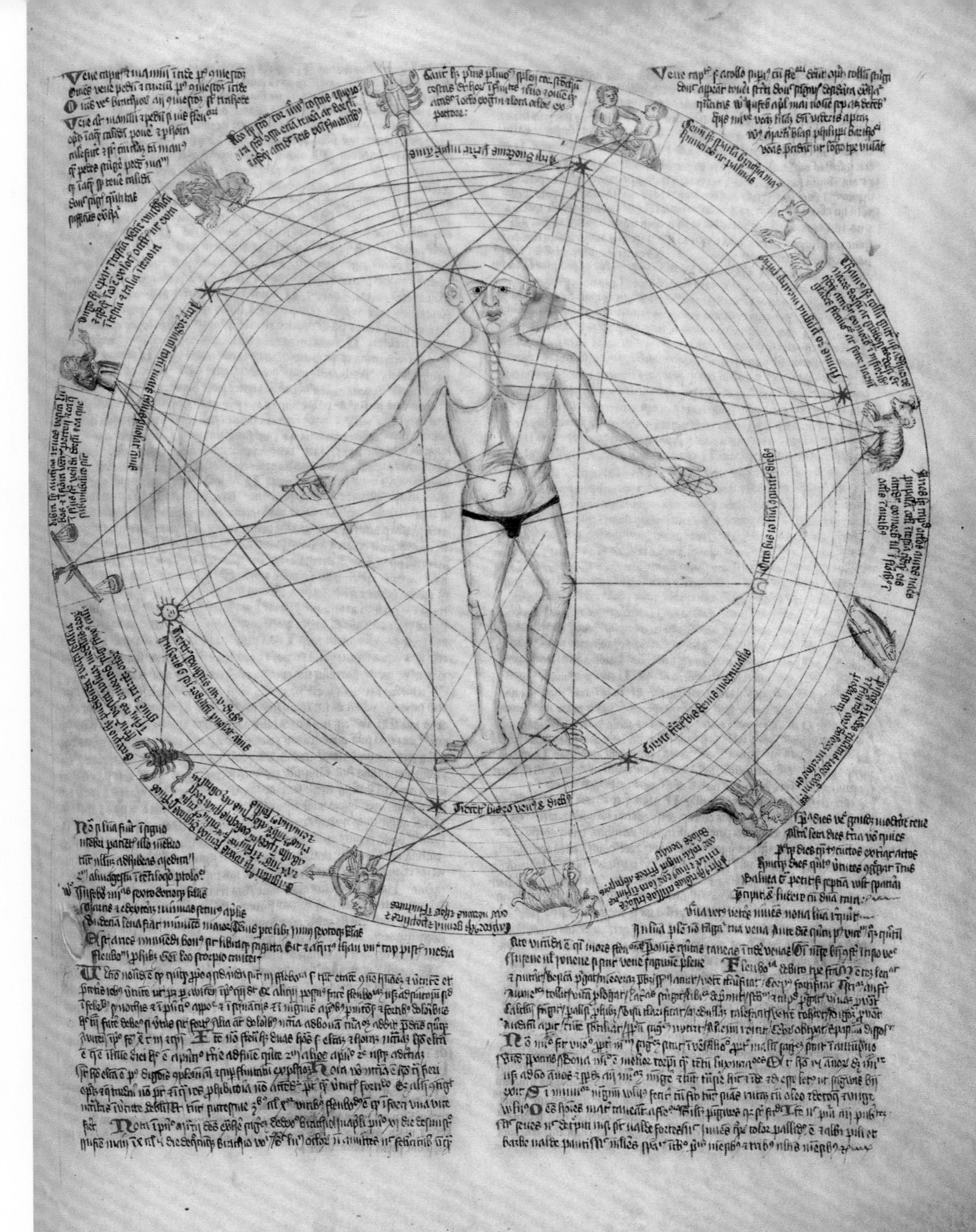

Physicians and the four humours

At the top of the tree was the university-trained physician. His name indicated his status. *Phusis* is the Greek word for nature, and the physician was the one who understood nature, and knew how to use natural products to repair the ways in which human nature went awry. He was a theorist, dispensing his physic from principles based on the description of the body set out by Aristotle in the fourth century B.C.E., and then codified in the first century C.E. by Galen of Pergamon, surgeon to gladiators. This ancient view of human nature, which persisted until microscopic discoveries began to suggest alternative approaches, asserted that all illnesses, physical and mental, were the result of imbalances in the four humours. The humours were the four fluids which pervaded the human body, and the tendency of one kind to dominate in an individual produced four basic 'temperaments'.

- **The sanguine temperament**
  Good blood led to a cheerful, robust constitution.
- **The melancholic temperament**
  A dominance of black bile, or 'melancholia', led to a depressive, but often artistic, personality.
- **The choleric temperament**
  Yellow bile or choler resulted in a hot-headed, quarrelsome character.
- **The phlegmatic**
  The predominance of sluggish phlegm produced a personality to suit.

The underlying pattern was complicated in various ways - the time of year would encourage one humour over another, as would the difference between youth and age - but the university-trained physician was sufficiently *doctus* (i.e. learned) to prescribe accurately.

Prescriptions

Physicians' prescriptions were based on the fact that the four temperaments of human nature were the human version of the four elemental principles which, according to the classical Greek theorists, permeated all of creation.

Everything in the universe was composed of four qualities: water (cold and moist), earth (cold and dry), fire (dry and hot), air (hot and moist).

What the physician had to do was to recognise how a fundamental humoral imbalance was causing a specific illness, and prescribe herbal preparations to rectify the situation, perhaps by sympathy (or homeopathy), but most often by providing the opposite principle (or allopathy).

Right: The four humours in man and the four elements in nature were aspects of the same pattern in the universe.
(illustration by Linda Kay)

Sanguine

(like warm and moist air) was seen as a cheerful and healthy temperament.

Good blood led to a cheerful, robust constitution.

Melancholic

(like earth) was seen as a cold and dry temperament.

A dominance of black bile, or 'melancholia', led to a depressive, but often artistic, personality.

earth

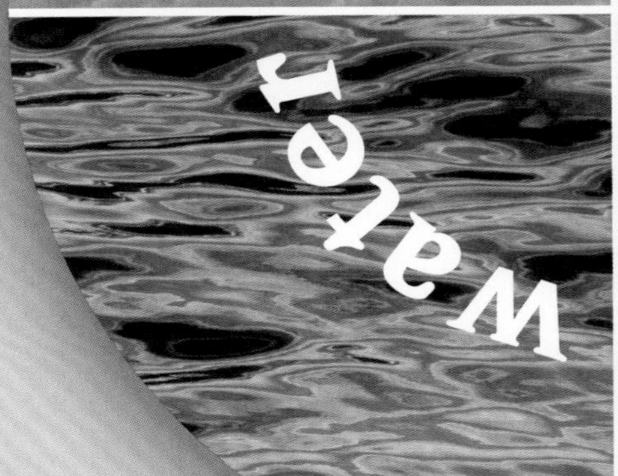

Phlegmatic

(like water) was seen as a cold and damp temperament.

The predominance of sluggish phlegm produced a personality to suit.

Choleric

(like fire) was seen as a hot and dry temperament.

Yellow bile or choler resulted in a hot-headed, quarrelsome character.

Herbals - required reading

Since all plants partook of nature's four principles, and since plants dominated medicine (though some minerals were used, as were parts of animals including their dung), every physician required a handbook - a herbal - which would identify plants so that the apothecary could either import them or buy them from his suppliers of local simples. A good herbal linked plants to specific illnesses and their cures, and noted where each species fitted in the hot/cold/dry/wet spectrum. Turner realised that most of his fellow-practitioners in England did not actually understand this full theoretical framework. At the end of his life, he introduced the last part of his *Herball* by spelling out what is meant by the 'degrees of herbs':

> '... there are certain herbs that are temperate... and are neither notably hot nor cold. And if any herb depart from the temperate herbs towards heat ... it is called hot in the first degree... If it be so hot as it can be, then it is called hot in the fourth degree.' (1568)

That is, the higher the degree the more dangerous the plant: 'medicines that are cold in the fourth degree', for instance, 'quench the natural heat' so completely that they 'kill men if they take them in great quantity'. One such is 'black poppy'. Horned poppy is both cold and dry in the fourth degree, so it too would kill in strong doses. Hence a herbal not only included a list of the illnesses and imbalances any plant could be used to treat, it also provided a quick reminder of how strong a plant was and whether it should be used only in moderate doses.

Left upper: *Veratrum album*, listed by Turner and his contemporaries as a kind of hellebore, was one of the deadly cold herbs. (photograph: Marie Addyman)

Left lower: *Papaver rhoeas*, the familiar red poppy or 'cornrose', which Turner describes in his carefully detailed discussion of different kinds of poppies and their several virtues and dangers. (photograph: Linda Kay)

Lavender species were used regularly in medicine and hygiene, but there was still some doubt as to how they related to other plants botanically.
(MS Ashmole 1504, fol.12r, courtesy of the Bodleian Library, University of Oxford)

The need for accurate plant naming

The main requirements of any herbal were to give the name of a plant, a recognisable description of its appearance, and a list of its applications. The problem was, however, that it was not at all clear that one writer was always referring to the same plant as another. The common core of reputable plant knowledge derived mainly from the Greek army doctor Dioscorides, who travelled into Asia and Northern Africa as well as in Southern Europe, and who wrote his *Materia Medica* in the first century C.E. Although this work was still revered in Turner's time, the flora that Dioscorides described was often very different from that seen by a Northern European. Furthermore, over the last half-century, hundreds of new plants had come in from the Americas. It followed that the names used in the first century might not apply to plants a sixteenth-century physician and his apothecary were prescribing under the same label.

The limited use of previous herbals

The confusion over the naming of plants was further confounded by the fact that the earliest extant text of Dioscorides is a manuscript of the sixth century. Hand-written manuscripts not only get lost and are easily destroyed, but they depend on the accuracy of the copyist - and repeated hand copying always produces errors. Since wars, plagues and other upheavals continually disrupted scholarly life, by the end of the Middle Ages herbals were often highly decorative, but as practical guides they were extremely limited. Medicine suffered, and therefore so did botany, since plant recognition was still primarily concerned with plant use.

However, winds of change were blowing through sixteenth-century learning, sometimes as the unexpected result of larger political events. When Constantinople fell to the Turks in 1453, some scholars fled westwards, taking both their knowledge of Greek and their Greek manuscripts with them. Turner, like others of his generation had access not only to Latin, the lingua franca of the international political, diplomatic and scholarly scene, but also to Greek, and to scholarly commentaries in contemporary European languages.

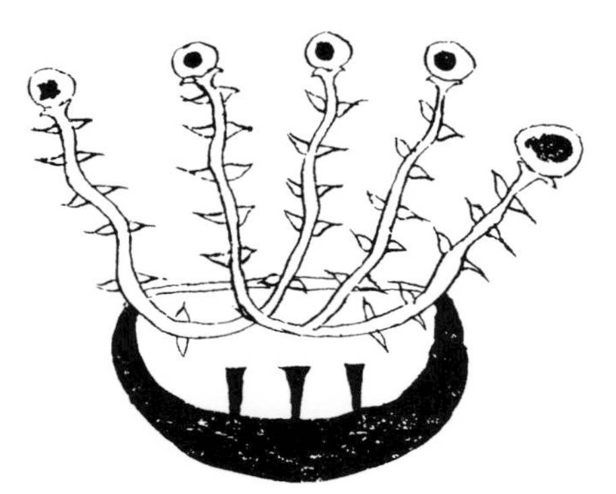

Left: This ancient image, supposed to represent saxifrage, shows how diagrammatic and unrecognisable illustrations had become before the work of the later naturalists.
(*Pseudo Apuleius MS*, after F.W.T. Hunger [1935])

Right: 'Lily' was a word which covered Iris and other genera in the 16th C, as well as the Madonna Lily shown here.
(MS Ashmole, 1504 fol.12r, courtesy of the Bodleian Library, University of Oxford)

Of the herbe one berrye.

Uchſius taught vs that the herbe that I call one berry/to be Aconitum pardalianches/ and then he thought it had ben ſo/and if he had kno wen a better/he wold haue ſhewed vs it. But Matthiolus proueth that the herbe whiche Fuchſius ſetteth furth for Aconito pardalian= che/is herba paris of yͤ later writers. The herb that I call One berrye/ hath a rounde ſtalke/ which is neuer aboue a ſpan long/ and oute of the middes therof commeth oute foure leaues/ not vnlyke vnto ſome Plantayne/ and in yͤ top of the ſtalke about a rounde black berrye come oute other foure ſmal leaues/and there in is ſede in color white. The roote is full of ſmall thinges / like thredes: This herbe groweth plentuouſlye in a wode beſyde Morpeth/called Cottinge woode/and in manye other woddes

An example of gothic black letter type from Turner's *Herball*. German books were still using this type in the 19th C, but English books by and large switched to italic by the end of the 16th C.

(Castle Morpeth Borough Council)

Type replaces calligraphy

With the development of movable type in printing, numerous works of reference became available. Printer-publishers realised increasingly that there was money to be made by creating large print-runs of compact editions called 'quartos' and 'octavos' for the international academic community. Now anyone anywhere could look at the same text, the same page, the same sentence, and compare notes. Initially books were printed in the old Gothic black-letter script, but later a new clear italic came to be used, as seen in Gerard's *Herball* of 1597.

Continental developments

All the developments in botanical scholarship would be experienced by Turner during his visit to the Continent. In Italy, he studied under Luca Ghini of Bologna and probably saw the first herbariums - collections of dried plants - being established. But it was in the German lands that several new important works were produced, in the herbals by Otto Brunfels, Hieronymus Bock, and Leonart Fuchs (for whom the genus *Fuchsia* is named). Each one made a crucial contribution to the development of medical botany. Brunfels included plants drawn from nature by a commissioned artist, and compared plants referred to by Dioscorides with those growing round Strasbourg. Bock emphasised the habitats of plants, and attempted to classify them under the headings of herbaceous plants, shrubs and trees. Fuchs' *De Historia Stirpium* of 1542 included about 100 native plants. A further breakthrough came in his German edition of 1543, which also gave a glossary of technical terms, and incorporated Bock's later research.

Accurate illustrations

It was Fuchs' woodcut illustrations of individual plants which particularly caught the eye. Already in some later mediaeval manuscript herbals there could be found plant portraits which clearly showed that the illustrator had looked at an actual plant. In the sixteenth century, printers commissioned an artist to create a drawing or painting of each plant, from which was created the woodcut. Woodcuts - or later, engravings - not only provided clear illustrations; once set up, they could be re-used time and again. Those in Fuchs' herbal were copied and borrowed repeatedly by writers of different nationalities, including Turner. The illustrations for his *Herball* came from one of the greatest pieces of botanical scholarship the sixteenth century had produced.

All these factors contributed to improvements in plant identification. As a flood of previously unknown specimens came in from the Americas; as the European flora was more carefully scrutinised; as newly available editions of ancient texts distinguished Dioscorides' plants from those in other regions; as better descriptions and illustrations were provided; so botany, conceived still as medical botany, flourished, even though the underlying principles of medical theory were slower to change.

Left: The illustrators of Fuchs' herbal.
(courtesy of the Wellcome Library, London)

Turner's aims and achievments

Over the thirty year period which began with the *Libellus* in 1538 and ended with the final part of the *Herball* in 1568, Turner's aims were clearly stated in the preface of each work. They were always intended to correct the major shortcomings of medical practice in his lifetime:

- There is a basic lack of knowledge about which herbs are which, and whether those collected in England are the same as those of the ancient text-books. Nobody can say confidently which local name applies to which plant in another locality, and certainly not how local designations square up to Latin scientific, names.
- Older herbals are garbled and add to the confusion.
- There are knowledgeable men in England who could provide clear and detailed text-books to clarify this muddle, but they do not do so, through lack of time or inclination.
- The result of all this is that physic in England is in dire straits, riddled with incompetence and ignorance throughout its entire hierarchy of practitioners and at risk of doing more harm than good in its blundering administration of prescribed medicines.

Turner therefore took upon himself the task of rectifying the blunders of misidentification by providing a succession of works in which medical plant material would be identified as accurately as possible. This he saw as both a Christian duty, undertaken on behalf of 'sicke folke' (1548), and a patriotic one. In the 1540s, studying the research prevalent in Italy and Germany, his patriotism took the form of a desire to show to the 'great honour of our country what number of sovereign and strange herbs were in England (and) not in other nations' (1548), going on to say that he hoped to describe 'every herb' in the country (1551). To establish England's medical-botanical status on the international stage would really necessitate him writing in Latin, and this he described as his first intention.

Above: Iris pseudacorus (which is probably the true fleur de lys) grows wild in Northern England and Scotland. It was classed as a lily in Turner's time.
(photograph: Marie Addyman)

Below: The daffodil, which Turner studied both from text books and from local knowledge. The name 'daffodil' and its variants is a corruption of 'affodil' or 'asphodel'.
(photograph: Linda Kay)

Turner chooses English

Given the dire state of medicine at home, what came to seem most important was to provide something accessible for his own fellow-countrymen, whose lamentable state of ignorance required him to write in English. After 1538 Turner consistently opted for his native language, defending his choice with the reminder that the founders of western medicine, Dioscorides and Galen, wrote in theirs, and that many of his European contemporaries, including the great Fuchs, had followed suit (1551).

Attacks on a herbal in English, according to Turner, rest primarily on the accusation that it would allow 'every man (and) every old wife (to) presume, not without the murder of many, to practice Physic'. Turner's blunt rejoinder is that, given that most apothecaries and physicians themselves couldn't actually understand the Latin texts on which their prescriptions depended, it was rather their 'dishonesty' which put 'in jeopardy' the lives of those they treat (1551). His own attitude is remarkable for his age. He began modestly enough, dedicating the little Latin text of his early career to his fellow medical students - or 'studious youth', as he quaintly put it. He ended by dedicating the final part of his herbal to surgeons, acknowledging their contribution to medicine, rather than seeing them as rivals or inferiors. But in between, he stated that he wrote not only for ignorant official practitioners (be they physicians, surgeons, or apothecaries), but also for the 'common people' (1551). Such a decision flew in the face of the most solemn academic and professional conventions, and therefore he had to defend it in each new publication.

To the right worshipfull Felowship and Companye of Surgiones of the citye of London chefely/and to all other that practyse Surgery within England/Williā Turner sendeth greting in Christ Jesu.

After that I had set furth two partes of my Herbal/conteining those plantes and herbes whereof ÿ old writers haue written and made mention / because that I knew both by myne owne experience/ and by other mennis writinge / that the herbes found after the old writers tyme/ if they were knowen with their vertues/should be very necessary for the healing of many diseases/greuous sicknesses/ woundes/ sores / brusinges/ and breaking as well inward as outward. I thought I should do no small benefit unto my countre/if I wrote of those plantes. Whereof I haue gathered as many as I knowe/or at the lest as many as came to my remembrance/leuing the rest that I haue not intreated of/to be intreated of other that haue more leasure then I haue. For surely beyng so much vexed with sicknes/and occupyed with preaching/and the study of Diuinitye and exercise of discipline/I haue had but smal leasure to write Herballes. This smal booke I geue and dedicate unto you/not only because this part for the bignes that it hath/intreateth most largely and plenteously of simples that belong unto Surgery/but also because beyng amongest you in London all that I was aquaynted withall/ and namely Maister Wright / late Surgion to the Quenes highnes/so willing and desyrous to learne and know such herbes as were not throwly knowen in England at that tyme. If ye take this my poore present in good worth/I thinke that I haue bestowed my laboures well / and if I can perceyue this / it may be an occasion / that if God send me health/leasure and longer life/that I take some more paynes for your profit in some other matter.

The Lorde kepe you.

At Welles 1 5 6 4. The 24. day of June.

Turner presented his last volume of the *Herball* to the Surgeons of London, in generous recognition that these men were now of great importance to the medical profession.

Of Elecampane.

Enula. Campana

Elecampane also grows in the Turner Garden, Morpeth. This is the illustration from Turner's *Herball*.
(Castle Morpeth Borough Council)

Early plant usage

Turner's open-minded attitude to medical practitioners seems far more individual and important than the actual details of herbal practice found in the *Herball*. It's true that probably all doctors had their own tips, favourite recipes, and individual areas of expertise, so that the recovery of Turner's 'Commonplace' books might give information about his individual use of herbs. Nevertheless the prescriptive use of plants in the *Herball* is interesting mainly for historical reasons, because it gives a clear indication of how they were generally used at this period.

It reminds us of herbs we still use, like comfrey for bones or fennel for digestion. It reminds us of those we now use as food (like onion - the juice of which Turner recommended to cure baldness), or in the flower garden (like lily of the valley), or those we've dismissed as weeds (like coltsfoot). It reminds us of the early categorisation of diseases and their strange names - squinancy, milt, St Anthony's fire, 'whitehaw or pearl in the eye'. It reminds us of anxieties we no longer share: Turner, like all his contemporaries dependent on classical herb lore, devotes a great deal of time to plants which supposedly would cure snake-bite. Pennyroyal, parsley, plantain, elecampane, bistort and chicory are but a few which mitigate 'the venom of serpents'.

Bistort (*Persicaria bistorta*) growing in the Turner Garden at Morpeth. Wild forms grow in various places in Northumberland.
(photograph: Emma Evans)

The father of English botany

Even though modern herbalism retains some early-modern plant uses, the underlying rationale and the whole concept of how the body functions have changed so dramatically that Turner's medical information is now largely a historical curiosity. Without this framework, we would not have the mature Turner's plant research, developing those studies he began as a boy observing the flora and fauna of Northumberland. Good physic is something he believes in passionately, but what gives him his title of 'the father of English botany' are those sections of his work where botany escapes from medicine, and a new kind of plant analysis is able to take shape. If Turner is radical in his insistence on making high-quality medical herbalism available to all who can benefit from it, in the field of early botanical identification he is indispensable.

Trailblazing work

Much of what Turner achieved, he did unaided. Had he lived a hundred years later, in the 1630s, he could have been part of a keen group of naturalists, who went on botanising expeditions, compared research and networked regularly. But he had few live models in his own country, and fewer written handbooks. *The Grete Herball* (1526) had incorporated some English plants and their English names, but nobody before him had realised that the only reliable way to get an accurate list of medical plants for English physicians was to mark 'in what place of England every herb grows' (1551).

Paris quadrifolia, with the characteristic parity which gives the plant its name and Cottingwood Burn, near Morpeth, where Turner found and recorded this plant.
(photographs: Marie Addyman and Brian Harle)

Detailed observation

The honourable profession of physic which Turner followed: an early illustration of some of the Greek masters of medicine.
(courtesy of the Wellcome Library, London)

Moreover, apart from the labour and time involved in such a mammoth task - and Turner complains more than once that his efforts were adversely affected both by illness and by the demands of his job – only a limited descriptive botanical vocabulary was available in the mid sixteenth century. After the invention of the microscope a hundred years later, it would be possible to understand and describe in more detail the different parts of the flower-head, but in Turner's time the stress on medically useful parts emphasised leaves, and roots. The common tactic was to describe one plant as like another, so meadowsweet has leaves like agrimony, and one form of Sorbus has 'a fruit like a little pear' (1548). Even within these restraints, Turner produces some acute observations:

> '(The) ash tree (has) leaves like unto the broader bay leaves, but they are sharper and indented round about the edges. The whole little foot stalk that all the leaves grow on is a green herbish thing and not woodish, and upon that the leaves grow in a distinct order, a small space going between one another, and they grow each side of the little stalk by couples... after the manner of the sorb apple leaves do grow. The seed of the ash tree groweth in long things, like birds' tongues, which are called... in English ash keys because they hang in bunches after the manner of keys' (1562).

Specific plant names

For the herbalist or the botanist, it is not enough to be able to describe a plant. We also need to give it a name. Because we unthinkingly use a nomenclature based on genus and species, we can not only tell a fellow-gardener that the plant they're admiring is a *Geranium*, we can also say that it's a particular species – *G. sanguineum* – or a particular cultivar – *G.* 'Ann Folkard'. For Turner it was quite different, as he came to realise very quickly.

At the beginning of his medical career, the need to correlate a list of names with detailed field-work didn't occur to him. What he did in the *Libellus* was to list officinal Latin names and their vernacular synonyms, on the basis of one synonym per country. Only in a few cases, where we would recognise different species within a genus (different kinds of Artemisia, for example: *A. arbrotanum*, *A. absinthium*, *A. vulgaris* etc.) does he point that 'There are two/ three sorts of…' (a plant). It is ten years later, after his experience of the new learning on the Continent, and when he decided to start from the English plant-name, that the scale of the problem begins to be apparent. Then he sees that the crucial question, 'Are we talking about the same plant?' has no simple answer. Turner's dilemma resulted from what seem like two completely contradictory problems.

Two cranesbills: meadow cranesbill (*Geranium pratense*) and bloody cranesbill (*Geranium sanguineum*). Both grow wild in Northumberland.
(photographs: Brian Harle)

Bellis perennis, the common daisy, as illustrated in Turner's *Herball*.

(Castle Morpeth Borough Council)

Problems of identification

On the one hand, there were too few scientific or 'officinal' (i.e. those recognised by the apothecaries) names for too many plants. The Latin scientific names derive from the great classical texts of Dioscorides, Pliny and Galen. They represented a strictly Mediterranean flora, but they had, over the centuries, been applied to plants from Northern Europe, which actually may either be different species within the same genus, or may represent entirely different genera. Furthermore, in the sixteenth century plants representing the entirely new floras of Central America and the Indies were coming in increasing numbers into Europe.

On the other hand, there were too many local names for too few plants. It isn't realistic to offer just one vernacular synonym per country against a Latin name. While some plants didn't even have an English name, others had different ones from county to county. A daisy could be 'banewort' in Northumberland, but 'twelve disciples' in Somerset. Not only that, but the term 'bachelor's buttons' referred to over a dozen different species, according to local usage.

Inevitably then, Latin/scientific/officinal names and local ones very often simply didn't match up. The problem, as Turner came to see it, wasn't therefore simply about ignorance of Latin; essentially, it was about examining individual plants in detail and comparing them. Only then could one decide between the names.

A botanical solution

The solution to plant naming was primarily botanical, rather than medical. While Turner does sometimes discuss the medical diagnosis that plant A is right for illness B, his major concern is that everyone agrees what plant A is. This requires Turner to use methods which we would recognise as those of standard scientific research, but which were almost unknown in the England of his time, though they were beginning to be developed on the Continent. What he did was to adapt the herbal's usual system of writing two paragraphs for each plant. The second paragraph still gives the usages according to the humoral classification of degree (and if necessary, different medical recommendations), but the first now extends the former brief reference to Latin and national names into a descriptive and comparative analysis.

How this works can be seen in his description of the daisy:

> A Dasey is called in Latin Bellis, in Dutch *klein tzitlosten* or *monatblumele*, in French *des margarites* and *pasquettes*, of the herbaries *consolida minor* or *primula veris*. There are two kinds of Dases: one with a red flower which growth in the gardens, and another which growth abroad in every green and highway. The Northern men call this herb a banewort because it helpeth bones to knit again. The leaf of the Dasey is something long and towards the end, round, and there are small nicks in the borders or edges of the leaves. Pliny writeth that the Dasey hath fifty three and sometimes fifty five little leaves which go about the yellow knop. It appeareth that the double Daseys were not known in Pliny's time, which have a great deal more than Pliny maketh mention of. (1551)

Left upper: A single form of the daisy.
(photograph: J.R. Crellin)

Left lower: A double form of the daisy.
(photograph: Marie Addyman)

The initial letter in the *Herball* for *Euonymus europaeus*, the spindle tree, shown with its characteristic berries in the photograph below. (Castle Morpeth Borough Council / Linda Kay)

Identifying wild plants

Reading this, we cannot help but wonder how a sixteenth-century writer can securely identify any plant when, it seems, even the common daisy (*Bellis perennis*) can be equated 'by the herbaries' with a cowslip (*Primula veris*)! As it happens, this isn't a sixteenth-century problem, since in Turner's time the Latin description of the first little flower of Spring (*Ver*) was not yet securely attached to this relative of the primrose. Even an entry as short as this demonstrates Turner's skills:

- First of all, he has actually looked at the plant: he attempts to describe the typical head of the compositae; he notes that the name covers different, but related forms; and he makes what will become a standard distinction between wild plants and those which are part of a burgeoning garden flora. He doesn't say here whether the red form is a wild native variant or an import, though he will do so in other cases.
- Secondly, he indicates by the list of foreign vernacular names that he can tap into contemporary European research. He can not only relate foreign names to local English ones, but he can compare actual plants from abroad with English examples.
- Thirdly, he is able to consult both Latin and Greek authorities, so that he can make the all-important distinction between those plants known right through from classical times and those which, even if related, are yet distinct from those mentioned by classical authors (what we would call a different species from the same genus). The final part of his *Herball* consists entirely of plants not mentioned by 'old authors'.
- Finally, each entry is backed up by an excellent woodcut derived from Fuchs.

The illustration in the *Herball* of Borage (*Borago officinalis*), the subject of one of Turner's most careful pieces of analysis.
(Castle Morpeth Borough Council)

Scholarly research

Turner is a model of scholarly research. He investigates both written and oral sources. He increases the likelihood of offering accurate texts by continually offering external evidence from his field work. He replaces second-hand, generalised hearsay with specific context and source wherever possible. Just as important, he defines his field of enquiry, which acknowledges what he is not attempting to do; and he presents his work as part of ongoing and co-operative research. It doesn't matter that Turner makes mistakes – as he does, for instance, in thinking of cotton lavender (*Santolina*) as a form of *Artemisia*. His mistakes are intelligent ones; they represent the considered state of his knowledge at the time; and he makes it clear that he is willing to correct them if he gains, or someone else supplies, better knowledge. Turner is one of the first English scientific writers in any field to recognise the importance of what he doesn't know.

'Every herb' in England

Scholarly working methods sustain Turner's pioneering efforts at recording English plants. Although he did not succeed in his aim of listing 'every herb' in England, it is estimated that he provides first records for about three hundred over his working life. The holly and the ivy come out of mediaeval carols to stand beside other trees and shrubs such as elm, privet, and yew. Several ferns are listed, including hart's tongue, lady's fern, and horse-tail. We find heather, dead nettle, foxglove, cowslip, fennel, pennyroyal, pansy, houseleek, and horned poppy. Often he mentions habitat: 'rocks and cliffs' as well as woods and 'water-sides'. References to Northumberland, to Cambridge, to the Thames near Syon, and to the countryside round Wells remind us of where he lived at different times and therefore where his searches could be most intensive. As his field-work developed he could recognise not only the same plant under its local names, but he began to perceive elementary groupings of different specimens, correcting his own earlier false impressions. At first, he thought the white-flowered plant popularly called *Laus tibi* was unrelated to the yellow one called 'daffadilly', but later he realised both were species of *Narcissus*.

Furthermore, to back up his descriptions, Turner tried to find a single name for each plant which could be generally acceptable. He made a start on this at the end of *The Names of herbs* by appending an alphabetical list of 'common English names used now in all counties of England'. This task, which was crucial if he was to achieve his aim of promoting the safe and uniform administration of herbal remedies by his fellow-countrymen, was complicated. Some plants, although native to England, nevertheless didn't have English names, so Turner simply invented them. He did this sometimes by adapting from another European language: monkshood (*Aconitum napellus*) and spindle tree (*Euonymus europaeus*) come from the Dutch. Elsewhere his classical

The foxglove, described for the first time in English records by Turner.
(photographs: Linda Kay)

training supplies him with a literal translation: loose strife, from the Greek *lysimachia*, is used to describe both yellow-flowered *lysimachia vulgaris* and purple-flowered *Lythrum salicaria*. Even though yellow and purple forms are now recognised as belonging to different genera, Turner's names have remained in each case. Some have dropped out of common use - bluebell is now more common than crowfoot - but honeysuckle, foxglove, herb robert and lady's mantle designate for us the very same plants that Turner observed and recorded five hundred years ago.

A timeless and familiar scene: English bluebells in Borough Woods, Morpeth.
(photographs: Linda Kay and Brian Harle)

I know no other use of the wilde ashe but that it is good to make cupbardes/ tables/ spoones & cuppes of. And that som use to make dagger hefters of the roote of it/ for it can scarsly be knowen from dudgyon/ and I thynke that the moste parte of dogion is of the root of the wilde ashe. Whatsoeuer uertue the other ashe hath thys must haue the same & more effectually/ sauyng in such maters as more moysture is requyred in. For then the comõ ashe is more fit for suche purposes.

Of Orobanche.

ORobanche/ as Dioscorides writeth/ is a redishe stalk two spannes hyghe/ and som tymes hygher/ tendre/ roughe without any lefe/ hath/ with a flour somthyng whitishe/ but turning toward yelow. The roote is a fynger thick. And when the stalk shrynkethe for dryues/ it is like an holow pype. It is playn that thys herbe groweth amonge certayn pulses/ & that it choketh & strangleth them/ where of it hath the name of Orobanche/ that is chokefitche or strangletare. Thus far Dioscorides of Orobanche. The herbe whiche I haue taken and taught xv. yeres ago to be Orobanche/ which also now of late yeares Matthiolus hath set out for Orobanche/ groweth in many places of Englãd/ bothe in the Northe countre besyde Morpethe/ whereas it is called our lady of new chapellis flour/ and also in the South countre a lytle from Shene in the broum closes. But it hath no name there. I haue sene it in diuerse places of Germany/ and first of all betwene Colon and Rodekirch. The herbe is comenly a fout long and oft longer/ I haue marked it many yeres/ but I colde neuer se any lefe upon it. But I haue sene the floures in diuerse places of diuerse colores/ and for the moste parte where so euer I saw thẽ/ they were redishe or turnyng to a purple color in som places/ but in figure they were lyke unto to ye floures of Clare with a thyng in them represẽtyng a cockis hede. The roote is round and much after the fasshon of a grete lekis hede/ and there grow out of it certayn long thynges lyke strynges which haue in them in certayn places sharp thynges lyke tethe/ where with it claspeth and holdeth the roote that it strangleth. I haue found it oft tymes claspyng & holdyng meruellously soft the rootes of broũ/ so that they looked as they had ben bound foulde oft about with small wyre. And ones I found thys herbe growyng besyd the comon clauer or medow trifoly/ which was all wethered/ and when I had dygged up the roote of the trifoly to se what shoulde be the cause that all other clauers or trifolies about wer grene and freshe/ that that trifoly should be dede. I found the rootes of Orobanche fast clasped about ye rootes of the clauer/ which as I did playnly perceyue/ draw out all the natural moysture from the herbe that it should haue lyued with all/ and so killed it/ as yui and dodder in continuance of tyme do with the trees and herbes that they fould and wynde them selues about. They that holde that cuscuta or doder is Orobanche in Dioscorides/ ar far deceyued. For Orobanche is a stalk and not a lace as doder is. Orobanche is but a fout and an half long/ but the laces of dodder will be som tyme

iii. or iiii. foote long. Orobanche hath a rootẽ a fynger thik/ but there is none such in doder/ for ye shall hardly fynde any ryght root at al in doder. The stalk of Orobanche is hollow when it is withered/ but so is not the stalk or rather the lace of doder. The stalke of Orobanche is roughe/ but the lace of doder is very smothe. Wherefore they were very far ouersene which now of late haue writen that doder is Orobanche in Dioscorides. Som other without any cause haue of late put thys herbe which I take to be Orobanche/ amongest the kyndes of Satyrion.

The properties of Orobanche.

ORobanche which may well be called in our tong chokefiche or stranglewede/ is eten comonly in sallates/ raw or sodden after the maner of sperage. Orobanche as Galene writeth is colde and dry in the firste degre. Matthiolus sayethe that Orobanche is called in Italian lupa/ that is a wolfe and also herba tora/ that is herbe bull/ because that if a cow chanse to eate of it/ she rynneth streyght way after to the bull. But it that Matthiolus wryteth agaynst Theophrast/ because he sayeth that Orobanche kylleth Orobus and strangleth it with hys pressyng in/ or thrystyng together/ and that Orobanche killeth pulses only with hys presence/ pleaseth me not/ as a saying agaynst reson autorite and experience. It is agaynst reson that only the presence of Orobanche should kill pulses/ seyng it is no uenummus herbe/ when euen uenummus herbes kill not them amongest whome they grow except they touche them/ or be so thyk amongest them that they take the norishmẽt from them/ whereby they should lyue. It is also both agaynst the autorite of Theophrast/ no lying writer/ and of lat agaynst Dioscorides/ whome he taketh in hand to expounde. For Dioscorides sayeth. It is playn that Orobanche groweth amongest pulses/ and that it chowketh or strangleth them/ where upon it hath gotten the name Orobanche/ that is Orobstrangler. Now I pray yow how can Orobanche strangle it that it toucheth not? Belyke Matthiolus saw no leues in Orobanche nor any claspers aboue the grounde/ & therefore he thought that there was no other thyng that Orobanche had/ where with it colde strangle/ & neuer marked ye litle stringes in the roote/ whiche not with out a faut hys Orobanche wanteth/ and so cam into thys error that Orobanche strangled only with hys presence. Tragus paynteth well Orobanch under the name of Satyrypnoni/ with such lytle strynges as it killed herbes with. And as touchynge experience/ I know that the freshe and yong Orobanche hath commyng out of the great roote/ many lytle strynges such as we se in a phrone or se lerr/ but longer/ wherewith it taketh holde of the rootes of the herbes that grow next unto it. Wherefore Matthiolus ought not so lyghtly to haue defaced the autorite of Theophrast so ancient and substantiall autor/ with laying ignorance unto hys charge/ seyng that Theophrastus in the same place where he speaketh of Orobanche telleth playnly that sum herbes ar first strangled by the roote/ and that not the only presence of suche wedes kill herbes and pulses/ but the takyng away of theyr norishement that

Conclusion

Broomrape (*Orobanche minor*) which was discovered by Turner at Our Lady's Chapel and Turner's *Herball* showing part of the entry for Orobanche.
(Castle Morpeth Borough Council / Brian Harle)

Turner poured all his passionate commitment and knowledge about plants into his books, which serve still as the indispensable first records for so much information concerning our native plants which we take for granted. He also provided lucid comparisons with as much of the European flora as he had encountered, thereby extending the range of herbal remedies for his contemporaries, and giving us a historical perspective for several of our garden plants.

Not only did he provide details about medicinal plants, he supported his medical practice with his late publications on various other aspects of health and hygiene; and he complemented his natural history with observations and records about both birds and fishes.

Throughout, Turner's aim was to be as accurate and as unbiased as possible. He wrote as clearly and precisely as he could; he corrected his mistakes; and he consulted as many comparative works as he had available. Unyielding in his belief that his countrymen needed and deserved better material than they had available, he was yet modest in the furtherance of his aims. Moreover, for most of the time, he was struggling against adversity of one kind or another - exile, poverty, ill-health, - and trying also to practice religious beliefs which frequently put him into danger.

Five hundred years after his birth, William Turner is at last being celebrated. Knowledge about him is moving out of specialist fields and becoming more generally accessible, allowing twenty-first century plant-lovers to appreciate just how much we owe to this zealous Tudor scholar.

Further reading and information

Turner's writings on plants

Libellus de Re Herbaria Novus (1538) and *The Names of Herbs* (1548); facsimile, ed. J. Britten, B. D. Jackson and W. T. Stearn for The Ray Society (London, 1965).

Libellus de Re Herbaria Novus (1538); ed. M. Ryden, H. Helander, K. Olsson (Uppsala, 1999).

A New Herball (Part 1, 1551); ed. G. Chapman and M. Tweddle (Mid-Northumberland Arts, 1989)

A New Herball (Parts 2 and 3, 1568); ed. G. Chapman, F. McCombie, A. U. Wesencraft (Cambridge, 1995)

Secondary sources

Jones, W. R. D., *William Turner, Tudor Naturalist, Physician and Divine* (London & New York, 1988)

Arber, A., *Herbals* (1912; reprinted Cambridge 1986)

Raven, C., *English Naturalists from Neckham to Ray* (1947, reprinted Cambridge & New York 1968)

Early-modern horticulture, botany and medicine

There is a great deal of information in print on these subjects. Good starting points would be, respectively:

Strong, R., *The Renaissance Garden in England* (London, 1979)

Pavord, A., *The Naming of Names* (London, 2005)

Siraisi, N., *Mediaeval and Early Renaissance Medicine* (Chicago and London, 1990)

The Reformation

To understand how the turmoil of the Reformation affected life in England:

Dickens, A.G., *The English Reformation* (London, rev.ed., 1967)

The Old Deanery Garden at Wells, situated under the Tudor windows of the Old Deanery.
(photograph: Marianne Adams)

Turner gardens

Castle Morpeth Borough Council has created
The William Turner Garden in Carlisle Park, open to the public every day.

www.castlemorpeth.gov.uk/carlislepark

The Old Deanery Garden at Wells holds open days and events each year.

www.olddeanerygarden.org.uk

Turner's *Herball*

Morpeth Chantry Bagpipe Museum displays Part I of Turner's *Herball* to the public.

amoore@castlemorpeth.gov.uk

The William Turner Garden in Carlisle Park, Morpeth, where can be found many of the plants used and recorded by Turner, son of this town.
(photograph: Castle Morpeth Borough Council)

Acknowledgements

The text of this booklet was prepared by M.E. Addyman for Castle Morpeth Borough Council,

The following have contributed photographs:
Marianne Adams
M.E. Addyman
J.R. Crellin
A.W. Davison
E. Evans
B. Harle
L.M. Kay
Richard Neale OBE
R.A. Singleton.

We are grateful to the Wellcome Trust, the National Portrait Gallery, the Bodleian Library, the Chapter of Wells and Castle Morpeth Borough Council for permission to use images and/or photographs.

The cover images include MS Ashmole 1504 fol.7v, courtesy of the Bodleian Library, University of Oxford.

The painting of the robin (page 4) was created specially by John Caffrey.

Publication was made possible through the support of Castle Morpeth Borough Council Community Fund and the Friends of Carlisle Park.

Design and artwork by
Linda K. Graphic Design Studio
0191 258 3464
mail@LindaK.net

Printed by Printers Coast Ltd
0191 296 4567
sales@printerscoast.co.uk